Houghton
Mifflin
Harcourt

W9-AXF-180

Made in the United States
Text printed on 100%
recycled paper

Houghton Mifflin Harcourt

Printed in the U.S.A.

ISBN 978-0-544-34215-6

20   0928   20

4500800120   C D E F G

Dear Students and Families,

Welcome to **Go Math!**, Grade 3! In this exciting mathematics program, there are hands-on activities to do and real-world problems to solve. Best of all, you will write your ideas and answers right in your book. In **Go Math!**, writing and drawing on the pages helps you think deeply about what you are learning, and you will really understand math!

By the way, all of the pages in your **Go Math!** book are made using recycled paper. We wanted you to know that you can Go Green with **Go Math!**

Sincerely,

The Authors

Made in the United States
Text printed on 100% recycled paper

GO MATH!

Authors

Juli K. Dixon, Ph.D.
Professor, Mathematics Education
University of Central Florida
Orlando, Florida

Edward B. Burger, Ph.D.
President, Southwestern University
Georgetown, Texas

Steven J. Leinwand
Principal Research Analyst
American Institutes for
  Research (AIR)
Washington, D.C.

Contributor

Rena Petrello
Professor, Mathematics
Moorpark College
Moorpark, California

Matthew R. Larson, Ph.D.
K-12 Curriculum Specialist for
  Mathematics
Lincoln Public Schools
Lincoln, Nebraska

Martha E. Sandoval-Martinez
Math Instructor
El Camino College
Torrance, California

English Language Learners Consultant

Elizabeth Jiménez
CEO, GEMAS Consulting
Professional Expert on English
  Learner Education
Bilingual Education and
  Dual Language
Pomona, California

Fractions

 Critical Area Developing understanding of fractions, especially unit fractions (fractions with numerator 1)

Critical Area

GO DIGITAL

Go online! Your math lessons are interactive. Use *i*Tools, Animated Math Models, the Multimedia eGlossary, and more.

Essential Question

What are equal parts of a whole?

Start

Chapter 9 Overview

In this chapter, you will explore and discover answers to the following **Essential Questions**:

• How can you compare fractions?

• What models can help you compare and order fractions?

• How can you use the size of the pieces to help you compare and order fractions?

• How can you find equivalent fractions?

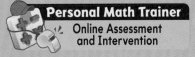

Personal Math Trainer
Online Assessment and Intervention

CRITICAL AREA REVIEW PROJECT THE SKATEBOARD DESIGNER: *www.thinkcentral.com*

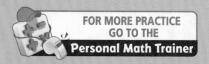

FOR MORE PRACTICE
GO TO THE
Personal Math Trainer

**Practice and
Homework**

Lesson Check and
Spiral Review in
every lesson

Compare Fractions

✔ **Show What You Know**

Check your understanding of important skills.

Name _____

▶ **Halves and Fourths** (1.G.A.3)

1. Find the shape that is divided into 2 equal parts. Color $\frac{1}{2}$.

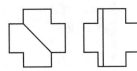

2. Find the shape that is divided into 4 equal parts. Color $\frac{1}{4}$.

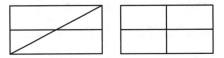

▶ **Parts of a Whole** Write the number of shaded parts and the number of equal parts. (2.G.A.2)

3. 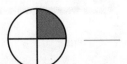 ____ shaded parts

____ equal parts

4. ____ shaded parts

____ equal parts

▶ **Fractions of a Whole**

Write the fraction that names the shaded part of each shape. (3.NF.A.1)

5.  _____

6. _____

7.  _____

Math in the Real World

Hannah keeps her marbles in bags with 4 marbles in each bag. She writes $\frac{3}{4}$ to show the number of red marbles in each bag. Find another fraction to name the number of red marbles in 2 bags.

Vocabulary Builder

▶ **Visualize It** ....

Complete the flow map by using the words with a ✓.

Fractions and Whole Numbers

What is it?		What are some examples?
_____	→	$\frac{2}{3} > \frac{1}{3}$
_____	→	$\frac{1}{4} < \frac{2}{4}$
_____	→	$\frac{1}{2} = \frac{2}{4}$
_____	→	$\frac{1}{3}, \frac{1}{4}$
_____	→	$\frac{2}{2}, \frac{4}{2}$

Review Words

compare
denominator
eighths
equal parts
equal to (=)
fourths
fraction
✓ greater than (>)
halves
✓ less than (<)
numerator
order
sixths
thirds
✓ unit fractions
✓ whole numbers

Preview Word

✓ equivalent
fractions

▶ **Understand Vocabulary** ....

Write the review word or preview word that answers the riddle.

1. We are two fractions that name the same amount.

   _____

2. I am the part of a fraction above the line. I tell how many parts are being counted.

   _____

3. I am the part of a fraction below the line. I tell how many equal parts are in the whole or in the group.

   _____

GO DIGITAL
• **Interactive Student Edition**
• **Multimedia eGlossary**

Chapter 9 Vocabulary

denominator

denominador

11

Eighths

octavos

17

Equal Parts

partes iguales

21

equivalent fractions

fracciones equivalentes

23

greater than (>)

mayor que

31

less than (<)

menor que

41

numerator

numerador

53

unit fraction

fracción unitaria

79

These are eighths

The part of a fraction below the line, which tells how many equal parts there are in the whole or in the group

Example: $\frac{1}{5}$ ←—— denominator

Two or more fractions that name the same amount

Example: $\frac{1}{2}$ and $\frac{3}{6}$ are equivalent fractions

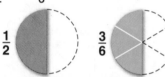

Parts that are exactly the same size

6 equal parts

A symbol used to compare two numbers when the lesser number is given first

Example:
Read 3 < 7 as "three is less than seven."

A symbol used to compare two numbers when the greater number is given first

Example:
Read 6 > 4 as "six is greater than four."

A fraction that has 1 as its top number, or numerator

Example: $\frac{1}{3}$ is a unit fraction

The part of a fraction above the line, which tells how many parts are being counted

Example: $\frac{1}{5}$ ←—— numerator

Pick It

For 3 players

Materials

• 4 sets of word cards

How to Play

1. Each player is dealt 5 cards. The remaining cards are a draw pile.

2. To take a turn, ask any player if he or she has a word that matches one of your word cards.

3. If the player has the word, he or she gives the card to you, and you must define the word.
   • If you are correct, keep the card and put the matching pair in front of you. Take another turn.
   • If you are wrong, return the card. Your turn is over.

4. If the player does not have the word, he or she answers, "Pick it." Then you take a card from the draw pile.

5. If the card you draw matches one of your word cards, follow the directions for Step 3 above. If it does not, your turn is over.

6. The game is over when one player has no cards left. The player with the most pairs wins.

Word Box
denominator
eighths
equal parts
equivalent fractions
greater than (>)
less than (<)
numerator
unit fractions

The Write Way

Reflect

Choose one idea. Write about it.

- Juan swam $\frac{3}{5}$ of a mile, and Greg swam $\frac{3}{8}$ of a mile. Explain how you know who swam farther.
- Explain how to compare two fractions.
- Write two examples of equivalent fractions and explain how you know they are equivalent.

Name _____

Problem Solving • Compare Fractions

Essential Question How can you use the strategy *act it out* to solve comparison problems?

 Common Core Number and Operations—Fractions—3.NF.A.3d *Also 3.NF.A.1*
MATHEMATICAL PRACTICES
MP2, MP4, MP5, MP6

Unlock the Problem

Mary and Vincent climbed up a rock wall at the park. Mary climbed $\frac{3}{4}$ of the way up the wall. Vincent climbed $\frac{3}{8}$ of the way up the wall. Who climbed higher?

You can act out the problem by using manipulatives to help you compare fractions.

Remember
$<$ is less than
$>$ is greater than
$=$ is equal to

Read the Problem	Solve the Problem
What do I need to find? _____ **What information do I need to use?** Mary climbed _____ of the way. Vincent climbed _____ of the way. **How will I use the information?** I will use _____ and _____ the lengths of the models to find who climbed _____ .	**Record the steps you used to solve the problem.**  Compare the lengths. ____ ◯ ____ The length of the $\frac{3}{4}$ model is _____ than the length of the $\frac{3}{8}$ model. So, _____ climbed higher on the rock wall.

Math Talk

MATHEMATICAL PRACTICES ④

Use Models When comparing fractions using fraction strips, how do you know which fraction is the lesser fraction?

🔑 Try Another Problem

Students at day camp are decorating paper circles for placemats. Tracy finished $\frac{3}{6}$ of her placemat. Kim finished $\frac{5}{6}$ of her placemat. Who finished more of her placemat?

Read the Problem	Solve the Problem
What do I need to find?	**Record the steps you used to solve the problem.**
What information do I need to use?	
How will I use the information?	

Math Talk

MATHEMATICAL PRACTICES ②

Use Reasoning How do you know that $\frac{5}{6}$ is greater than $\frac{3}{6}$ without using models?

1. How did your model help you solve the problem? _____

_____

2. Tracy and Kim each had a carton of milk with lunch. Tracy drank $\frac{5}{8}$ of her milk. Kim drank $\frac{7}{8}$ of her milk. Who drank more of her milk? Explain.

_____

Name _____

Unlock the Problem

✓ Circle the question.
✓ Underline important facts.
✓ Act out the problem using manipulatives.

✓ 1. At the park, people can climb a rope ladder to its top. Rosa climbed $\frac{2}{8}$ of the way up the ladder. Justin climbed $\frac{2}{6}$ of the way up the ladder. Who climbed higher on the rope ladder?

First, what are you asked to find?

_____

Then, model and compare the fractions. Think: Compare $\frac{2}{8}$ and $\frac{2}{6}$.

Last, find the greater fraction.

____ ◯ ____

So, _____ climbed higher on the rope ladder.

✓ 2. What if Cara also tried the rope ladder and climbed $\frac{2}{4}$ of the way up? Who climbed highest on the rope ladder: Rosa, Justin, or Cara? Explain how you know.

_____

_____

_____

On Your Own

3. (MATHEMATICAL PRACTICE ⑤) **Use a Concrete Model** Ted walked $\frac{2}{3}$ mile to his soccer game. Then he walked $\frac{1}{3}$ mile to his friend's house. Which distance is shorter? Explain how you know.

_____

_____

Use the table for 4–5.

4. **GO DEEPER** Suri is spreading jam on 8 biscuits for breakfast. The table shows the fraction of biscuits spread with each jam flavor. Which flavor did Suri use on the most biscuits?

   Hint: Use 8 counters to model the biscuits.

   _____

Suri's Biscuits	
Jam Flavor	**Fraction of Biscuits**
Peach	$\frac{3}{8}$
Raspberry	$\frac{4}{8}$
Strawberry	$\frac{1}{8}$

5. **WRITE ▸Math** **What's the Question?** The answer is strawberry.

   _____

   _____

WRITE ▸Math • Show Your Work

6. **THINK SMARTER** Suppose Suri had also used plum jam on the biscuits. She frosted $\frac{1}{2}$ of the biscuits with peach jam, $\frac{1}{4}$ with raspberry jam, $\frac{1}{8}$ with strawberry jam, and $\frac{1}{8}$ with plum jam. Which flavor of jam did Suri use on the most biscuits?

   _____

7. Ms. Gordon has many snack bar recipes. One recipe uses $\frac{1}{3}$ cup oatmeal, $\frac{1}{4}$ cup of milk, and $\frac{1}{2}$ cup flour. Which ingredient will Ms. Gordon use the most of?

   _____

8. **THINK SMARTER** Rick lives $\frac{4}{6}$ mile from school. Noah lives $\frac{3}{6}$ mile from school.

   Use the fractions and symbols to show which distance is longer.

   $\frac{3}{6}$, $\frac{4}{6}$, < and >   ☐ ○ ☐

Problem Solving • Compare Fractions

Common Core **COMMON CORE STANDARD—3.NF.A.3d**
Develop understanding of fractions as numbers.

Solve.

1. Luis skates $\frac{2}{3}$ mile from his home to school. Isabella skates $\frac{2}{4}$ mile to get to school. Who skates farther?

   Think: Use fraction strips to act it out.

   _____ Luis _____

2. Sandra makes a pizza. She puts mushrooms on $\frac{2}{8}$ of the pizza. She adds green peppers to $\frac{5}{8}$ of the pizza. Which topping covers more of the pizza?

   _____

3. The jars of paint in the art room have different amounts of paint. The green paint jar is $\frac{4}{8}$ full. The purple paint jar is $\frac{4}{6}$ full. Which paint jar is less full?

   _____

4. Jan has a recipe for bread. She uses $\frac{2}{3}$ cup of flour and $\frac{1}{3}$ cup of chopped onion. Which ingredient does she use more of, flour or onion?

   _____

5. WRITE ▶Math Explain how you can find whether $\frac{5}{6}$ or $\frac{5}{8}$ is greater.

   _____

   _____

Lesson Check (3.NF.A.3d)

1. Ali and Jonah collect seashells in identical buckets. When they are finished, Ali's bucket is $\frac{2}{6}$ full and Jonah's bucket is $\frac{3}{6}$ full. Compare the fractions using >, < or =.

$$\frac{3}{6} \bigcirc \frac{2}{6}$$

2. Rosa paints a wall in her bedroom. She puts green paint on $\frac{5}{8}$ of the wall and blue paint on $\frac{3}{8}$ of the wall. Compare the fractions using >, < or =.

$$\frac{5}{8} \bigcirc \frac{3}{8}$$

Spiral Review (3.OA.B.6, 3.OA.D.9, 3.NF.A.1)

3. Dan divides a pie into eighths. How many equal parts are there?

_____

4. Draw lines to divide the circle into 4 equal parts.

5. Charles places 30 pictures on his bulletin board in 6 equal rows. How many pictures are in each row?

6. Describe a pattern in the table.

Tables	1	2	3	4	5
Chairs	5	10	15	20	25

FOR MORE PRACTICE GO TO THE Personal Math Trainer

Name _____

Compare Fractions with the Same Denominator

Essential Question How can you compare fractions with the same denominator?

Common Core
Number and Operations—Fractions—
3.NF.A.3d Also 3.NF.A.1, 3.NF.A.2b
MATHEMATICAL PRACTICES
MP2, MP3, MP6

 Unlock the Problem

Jeremy and Christina are each making a quilt block. Both blocks are the same size and both are made of 4 equal-size squares. $\frac{2}{4}$ of Jeremy's squares are green. $\frac{1}{4}$ of Christina's squares are green. Whose quilt block has more green squares?

• Circle the two fractions you need to compare.

• How are the two fractions alike?

_____

🔒 **Compare fractions of a whole.**

• Shade $\frac{2}{4}$ of Jeremy's quilt block.

• Shade $\frac{1}{4}$ of Christina's quilt block.

• Compare $\frac{2}{4}$ and $\frac{1}{4}$.

The greater fraction will have the larger amount of the whole shaded.

$$\frac{2}{4} \bigcirc \frac{1}{4}$$

Jeremy's Quilt Block   **Christina's Quilt Block**

Math Idea

You can compare two fractions when they refer to the same whole or to groups that are the same size.

So, _____ quilt block has more green squares.

🔒 **Compare fractions of a group.**

Jen and Maggie each have 6 buttons.

• Shade 3 of Jen's buttons to show the number of buttons that are red. Shade 5 of Maggie's buttons to show the number that are red.

• Write a fraction to show the number of red buttons in each group. Compare the fractions.

Jen's Buttons

Maggie's Buttons

There are the same number of buttons in each group, so you can count the number of red buttons to compare the fractions.

$$3 < \underline{\quad}, \text{ so } \frac{\underline{}}{6} < \frac{\underline{}}{6}.$$

So, _____ has a greater fraction of red buttons.

Use fraction strips and a number line.

At the craft store, one piece of ribbon is $\frac{2}{8}$ yard long. Another piece of ribbon is $\frac{7}{8}$ yard long. If Sean wants to buy the longer piece of ribbon, which piece should he buy?

- On a number line, a fraction farther to the right is greater than a fraction to its left.
- On a number line, a fraction farther to the left is _____ a fraction to its right.

Compare $\frac{2}{8}$ and $\frac{7}{8}$.

- Shade the fraction strips to show the locations of $\frac{2}{8}$ and $\frac{7}{8}$.

- Draw and label points on the number line to represent the distances $\frac{2}{8}$ and $\frac{7}{8}$.

- Compare the lengths.

  $\frac{2}{8}$ is to the left of $\frac{7}{8}$. It is closer to $\frac{0}{8}$, or _____.

  $\frac{7}{8}$ is to the _____ of $\frac{2}{8}$. It is closer to ——, or _____.

  —— < —— and —— > ——

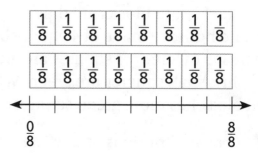

So, Sean should buy the piece of ribbon that is —— yard long.

Use reasoning.

Ana and Omar are decorating same-size bookmarks. Ana covers $\frac{3}{3}$ of her bookmark with glitter. Omar covers $\frac{1}{3}$ of his bookmark with glitter. Whose bookmark is covered with more glitter?

Compare $\frac{3}{3}$ and $\frac{1}{3}$.

- When the denominators are the same, the whole is divided into same-size pieces. You can look at the _____ to compare the number of pieces.

- Both fractions involve third-size pieces. _____ pieces are more than _____ piece. 3 > _____, so —— > ——.

So, _____ bookmark is covered with more glitter.

Math Talk

MATHEMATICAL PRACTICES ⑥

Explain how you can use reasoning to compare fractions with the same denominator.

Name _____

Reason Abstractly Why do fractions increase in size as you move right on the number line?

1. Draw points on the number line to show $\frac{1}{6}$ and $\frac{5}{6}$. Then compare the fractions.

$$\frac{0}{6} \quad \frac{1}{6} \quad \frac{2}{6} \quad \frac{3}{6} \quad \frac{4}{6} \quad \frac{5}{6} \quad \frac{6}{6}$$

Think: $\frac{1}{6}$ is to the left of $\frac{5}{6}$ on the number line.

$$\frac{1}{6} \bigcirc \frac{5}{6}$$

Compare. Write <, >, or =.

2. $\frac{4}{8} \bigcirc \frac{3}{8}$ ✓3. $\frac{1}{4} \bigcirc \frac{4}{4}$ 4. $\frac{1}{2} \bigcirc \frac{1}{2}$ ✓5. $\frac{3}{6} \bigcirc \frac{2}{6}$

On Your Own

Compare. Write <, >, or =.

6. $\frac{2}{4} \bigcirc \frac{3}{4}$ 7. $\frac{2}{3} \bigcirc \frac{2}{3}$ 8. $\frac{4}{6} \bigcirc \frac{2}{6}$ 9. $\frac{0}{8} \bigcirc \frac{2}{8}$

THINK SMARTER Write a fraction less than, greater than, or equal to the given fraction.

10. $\frac{1}{2} < \underline{\quad}$ 11. $\underline{\quad} < \frac{12}{6}$ 12. $\frac{8}{8} = \underline{\quad}$ 13. $\underline{\quad} > \frac{2}{4}$

Problem Solving • Applications

14. Carlos finished $\frac{5}{8}$ of his art project on Monday. Tyler finished $\frac{7}{8}$ of his art project on Monday. Who finished more of his art project on Monday?

15. **MATHEMATICAL PRACTICE ②** **Use Reasoning** Ms. Endo made two loaves of bread that are the same size. Her family ate $\frac{1}{4}$ of the banana bread and $\frac{3}{4}$ of the cinnamon bread. Which loaf of bread had less left over?

16. **THINK SMARTER** Todd and Lisa are comparing fraction strips. Which statements are correct? Mark all that apply.

   Ⓐ $\frac{1}{4} < \frac{4}{4}$ Ⓑ $\frac{5}{6} < \frac{4}{6}$ Ⓒ $\frac{2}{3} > \frac{1}{3}$ Ⓓ $\frac{5}{8} > \frac{4}{8}$

THINK SMARTER **What's the Error?**

17. Gary and Vanessa are comparing fractions. Vanessa models $\frac{2}{4}$ and Gary models $\frac{3}{4}$. Vanessa writes $\frac{3}{4} < \frac{2}{4}$. Look at Gary's model and Vanessa's model and describe her error.

Vanessa's Model

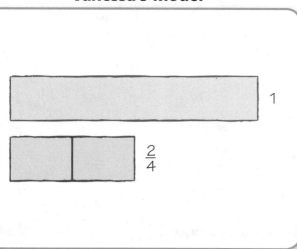

Gary's Model

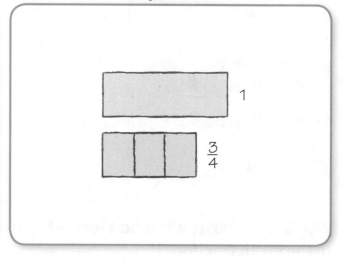

• Describe Vanessa's error.

_____

_____

18. **GO DEEPER** Explain how to correct Vanessa's error. Then show the correct model.

_____

_____

_____

_____

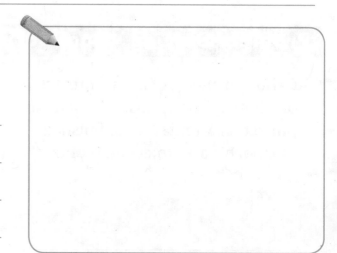

Compare Fractions with the Same Denominator

COMMON CORE STANDARD—3.NF.A.3d
Develop understanding of fractions as numbers.

Compare. Write <, >, or =.

1. $\frac{3}{4} \bigotimes{>} \frac{1}{4}$

2. $\frac{3}{6} \bigcirc \frac{0}{6}$

3. $\frac{1}{2} \bigcirc \frac{1}{2}$

4. $\frac{5}{6} \bigcirc \frac{6}{6}$

5. $\frac{7}{8} \bigcirc \frac{5}{8}$

6. $\frac{2}{3} \bigcirc \frac{3}{3}$

7. $\frac{8}{8} \bigcirc \frac{0}{8}$

8. $\frac{1}{6} \bigcirc \frac{1}{6}$

9. $\frac{3}{4} \bigcirc \frac{2}{4}$

10. $\frac{1}{6} \bigcirc \frac{2}{6}$

11. $\frac{1}{2} \bigcirc \frac{0}{2}$

12. $\frac{3}{8} \bigcirc \frac{3}{8}$

13. $\frac{1}{4} \bigcirc \frac{4}{4}$

14. $\frac{5}{8} \bigcirc \frac{4}{8}$

15. $\frac{4}{6} \bigcirc \frac{6}{6}$

Problem Solving Real World

16. Ben mowed $\frac{5}{6}$ of his lawn in one hour. John mowed $\frac{4}{6}$ of his lawn in one hour. Who mowed less of his lawn in one hour?

17. Darcy baked 8 muffins. She put blueberries in $\frac{5}{8}$ of the muffins. She put raspberries in $\frac{3}{8}$ of the muffins. Did more muffins have blueberries or raspberries?

_____

_____

18. **WRITE** ▸Math Explain how you can use reasoning to compare two fractions with the same denominator.

_____

_____

Lesson Check (3.NF.A.3d)

1. Julia paints $\frac{2}{6}$ of a wall in her room white. She paints more of the wall green than white. What fraction could show the part of the wall that is green?

_____

2. Compare. Write <, >, or =.

$$\frac{2}{8} \bigcirc \frac{3}{8}$$

Spiral Review (3.OA.A.3, 3.OA.B.5, 3.OA.C.7, 3.NBT.A.3)

3. Mr. Edwards buys 2 new knobs for each of his kitchen cabinets. The kitchen has 9 cabinets. How many knobs does he buy?

_____

4. Allie builds a new bookcase with 8 shelves. She can put 30 books on each shelf. How many books can the bookcase hold?

_____

5. The Good Morning Café has 28 customers for breakfast. There are 4 people sitting at each table. How many tables are filled?

_____

6. Ella wants to use the Commutative Property of Multiplication to help find the product 5×4. What number sentence can she use?

_____

FOR MORE PRACTICE
GO TO THE
Personal Math Trainer

Compare Fractions with the Same Numerator

Common Core — Number and Operations—Fractions—3.NF.A.3d *Also 3.NF.A.1*
MATHEMATICAL PRACTICES
MP1, MP2, MP6

Essential Question How can you compare fractions with the same numerator?

Unlock the Problem

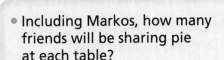

Markos is at Athena's Cafe. He can sit at a table with 5 of his friends or at a different table with 7 of his friends. The same-size spinach pie is shared equally among the people at each table. At which table should Markos sit to get more pie?

 Model the problem.

There will be 6 friends sharing Pie A or 8 friends sharing Pie B.

So, Markos will get either $\frac{1}{6}$ or $\frac{1}{8}$ of a pie.

- Shade $\frac{1}{6}$ of Pie A.
- Shade $\frac{1}{8}$ of Pie B.
- Which piece of pie is larger?
- Compare $\frac{1}{6}$ and $\frac{1}{8}$.

$$\frac{1}{6} \bigcirc \frac{1}{8}$$

So, Markos should sit at the table with _____ friends to get more pie.

- Including Markos, how many friends will be sharing pie at each table?

_____

- What will you compare?

_____

_____

Pie A   **Pie B**

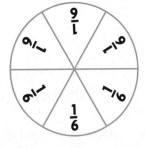   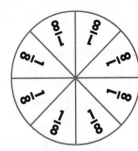

Math Talk

MATHEMATICAL PRACTICES ❶

Make Sense of Problems
Suppose Markos wants two pieces of one of the pies above. Is $\frac{2}{6}$ or $\frac{2}{8}$ of the pie a greater amount? Explain how you know.

1. Which pie has more pieces? _____
   The *more* pieces a whole is divided into,

   the _____ the pieces are.

2. Which pie has fewer pieces? _____
   The *fewer* pieces a whole is divided into,

   the _____ the pieces are.

Use fraction strips.

On Saturday, the campers paddled $\frac{2}{8}$ of their planned route down the river. On Sunday, they paddled $\frac{2}{3}$ of their route down the river. On which day did the campers paddle farther?

Compare $\frac{2}{8}$ and $\frac{2}{3}$.

- Place a ✓ next to the fraction strips that show more parts in the whole.

- Shade $\frac{2}{8}$. Then shade $\frac{2}{3}$. Compare the shaded parts.

- $\frac{2}{8} \bigcirc \frac{2}{3}$

1							
$\frac{1}{8}$	$\frac{1}{8}$	$\frac{1}{8}$	$\frac{1}{8}$	$\frac{1}{8}$	$\frac{1}{8}$	$\frac{1}{8}$	$\frac{1}{8}$

$\frac{1}{3}$	$\frac{1}{3}$	$\frac{1}{3}$

Think: $\frac{1}{8}$ is less than $\frac{1}{3}$, so $\frac{2}{8}$ is less than $\frac{2}{3}$.

So, the campers paddled farther on _____.

 Use reasoning.

For her class party, Felicia baked two trays of snacks that were the same size. After the party, she had $\frac{3}{4}$ of the carrot snack and $\frac{3}{6}$ of the apple snack left over. Was more carrot snack or more apple snack left over?

Compare $\frac{3}{4}$ and $\frac{3}{6}$.

- Since the numerators are the same, look at the denominators to compare the size of the pieces.

$\frac{3}{4} \bullet \frac{3}{6}$

> - The *more* pieces a whole is divided into, the _____ the pieces are.
> - The *fewer* pieces a whole is divided into, the _____ the pieces are.

- $\frac{1}{4}$ is _____ than $\frac{1}{6}$ because there are _____ pieces.

 ERROR Alert

When comparing fractions with the same numerator, be sure the symbol shows that the fraction with fewer pieces in the whole is the greater fraction.

- $\frac{3}{4} \bigcirc \frac{3}{6}$

So, there was more of the _____ snack left over.

Name _____

1. Shade the models to show $\frac{1}{6}$ and $\frac{1}{4}$.

Then compare the fractions.

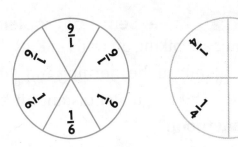

$\frac{1}{6} \bigcirc \frac{1}{4}$

Compare. Write <, >, or =.

✓**2.** $\frac{1}{8} \bigcirc \frac{1}{3}$ ✓**3.** $\frac{3}{4} \bigcirc \frac{3}{8}$ **4.** $\frac{2}{6} \bigcirc \frac{2}{3}$

5. $\frac{4}{8} \bigcirc \frac{4}{4}$ **6.** $\frac{3}{6} \bigcirc \frac{3}{6}$ **7.** $\frac{8}{4} \bigcirc \frac{8}{8}$

Math Talk MATHEMATICAL PRACTICES ①

Evaluate Why is $\frac{1}{2}$ greater than $\frac{1}{4}$?

On Your Own

Compare. Write <, >, or =.

8. $\frac{1}{3} \bigcirc \frac{1}{4}$ **9.** $\frac{2}{3} \bigcirc \frac{2}{6}$ **10.** $\frac{4}{8} \bigcirc \frac{4}{2}$

11. $\frac{6}{8} \bigcirc \frac{6}{6}$ **12.** $\frac{1}{6} \bigcirc \frac{1}{2}$ **13.** $\frac{7}{8} \bigcirc \frac{7}{8}$

14. **GO DEEPER** James ate $\frac{3}{4}$ of his quesadilla. David ate $\frac{2}{3}$ of his quesadilla. Both are the same size. Who ate more of his quesadilla?

James said he knows he ate more because he looked at the amounts left. Does his answer make sense? Shade the models. Explain.

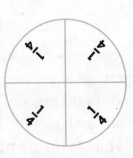

James   David

Unlock the Problem Real World

15. **MATHEMATICAL PRACTICE ①** **Make Sense of Problems** Quinton and Hunter are biking on trails in Katy Trail State Park. They biked $\frac{5}{6}$ mile in the morning and $\frac{5}{8}$ mile in the afternoon. Did they bike a greater distance in the morning or in the afternoon?

a. What do you need to know? _____

b. The numerator is 5 in both fractions, so compare $\frac{1}{6}$ and $\frac{1}{8}$. Explain.

_____

c. How can you solve the problem?

d. Complete the sentences.

In the morning, the boys biked

_____ mile. In the afternoon, they biked _____ mile.

So, the boys biked a greater distance

in the _____ . $\frac{5}{6}$ ◯ $\frac{5}{8}$

16. **THINK SMARTER** Zach has a piece of pie that is $\frac{1}{4}$ of a pie. Max has a piece of pie that is $\frac{1}{2}$ of a pie. Max's piece is smaller than Zach's piece. Explain how this could happen. Draw a picture to show your answer.

_____

_____

Personal Math Trainer

17. **THINK SMARTER +** Before taking a hike, Kate and Dylan each ate part of their same-size granola bars. Kate ate $\frac{1}{3}$ of her bar. Dylan ate $\frac{1}{2}$ of his bar. Who ate more of the granola bar? Explain how you solved the problem.

_____

_____

Name _____

Compare Fractions with the Same Numerator

COMMON CORE STANDARD—3.NF.A.3d
Develop understanding of fractions as numbers.

Compare. Write <, >, or =.

1. $\frac{1}{8}$ $<$ $\frac{1}{2}$

2. $\frac{3}{8}$ ◯ $\frac{3}{6}$

3. $\frac{2}{3}$ ◯ $\frac{2}{4}$

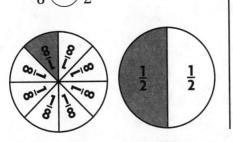

4. $\frac{2}{8}$ ◯ $\frac{2}{3}$

5. $\frac{3}{6}$ ◯ $\frac{3}{4}$

6. $\frac{1}{2}$ ◯ $\frac{1}{6}$

7. $\frac{5}{6}$ ◯ $\frac{5}{8}$

8. $\frac{4}{8}$ ◯ $\frac{4}{8}$

9. $\frac{6}{8}$ ◯ $\frac{6}{6}$

Problem Solving

10. Javier is buying food in the lunch line. The tray of salad plates is $\frac{3}{8}$ full. The tray of fruit plates is $\frac{3}{4}$ full. Which tray is more full?

11. Rachel bought some buttons. Of the buttons, $\frac{2}{4}$ are yellow and $\frac{2}{8}$ are red. Rachel bought more of which color buttons?

_____ _____

12. **WRITE** ▸*Math* Explain how the number of pieces in a whole relates to the size of each piece.

_____

_____

Lesson Check (3.NF.A.3d)

1. What symbol makes the statement true? Write <, >, or =.

$\frac{3}{4}$ $\frac{3}{8}$

2. What symbol makes the statement true? Write <, >, or =.

$\frac{2}{4}$ $\frac{2}{3}$

Spiral Review (3.OA.C.7, 3.NF.A.1)

3. Anita divided a circle into 6 equal parts and shaded 1 of the parts. What fraction names the part she shaded?

4. What fraction names the shaded part of the rectangle?

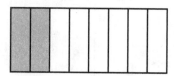

5. Chip worked at the animal shelter for 6 hours each week for several weeks. He worked for a total of 42 hours. How many weeks did Chip work at the animal shelter?

6. Mr. Jackson has 20 quarters. If he gives 4 quarters to each of his children, how many children does Mr. Jackson have?

© Houghton Mifflin Harcourt Publishing Company

FOR MORE PRACTICE
GO TO THE
Personal Math Trainer

Name _____

Compare Fractions

Essential Question What strategies can you use to compare fractions?

Number and Operations—
Fractions—3.NF.A.3d *Also 3.NF.A.1,
3.NF.A.3*
MATHEMATICAL PRACTICES
MP1, MP2, MP4, MP6

Unlock the Problem

Luka and Ann are eating the same-size small pizzas. One plate has $\frac{3}{4}$ of Luka's cheese pizza. Another plate has $\frac{5}{6}$ of Ann's mushroom pizza. Whose plate has more pizza?

 Compare $\frac{3}{4}$ and $\frac{5}{6}$.

Missing Pieces Strategy
• You can compare fractions by comparing pieces missing from a whole.

• Shade $\frac{3}{4}$ of Luka's pizza and $\frac{5}{6}$ of Ann's pizza. Each fraction represents a whole that is missing one piece.

• Since $\frac{1}{6}$ ◯ $\frac{1}{4}$, a smaller piece is missing from Ann's pizza.

• If a smaller piece is missing from Ann's pizza, she must have more pizza.

So, _____ plate has more pizza.

• Circle the numbers you need to compare.
• How many pieces make up each whole pizza?

_____

Luka   **Ann**

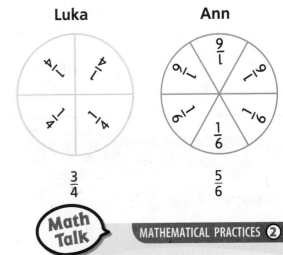

$\frac{3}{4}$ $\frac{5}{6}$

Math Talk   MATHEMATICAL PRACTICES ②

Reason Abstractly How does knowing that $\frac{1}{4}$ is less than $\frac{1}{3}$ help you compare $\frac{3}{4}$ and $\frac{2}{3}$?

Morgan ran $\frac{2}{3}$ mile. Alexa ran $\frac{1}{3}$ mile. Who ran farther?

 Compare $\frac{2}{3}$ and $\frac{1}{3}$.

$$\frac{}{3} > \frac{}{3}$$

Same Denominator Strategy
• When the denominators are the same, you can compare only the number of pieces, or the numerators.

So, _____ ran farther.

Ms. Davis is making a fruit salad with $\frac{3}{4}$ pound of cherries and $\frac{3}{8}$ pound of strawberries. Which weighs less, the cherries or the strawberries?

 Compare $\frac{3}{4}$ and $\frac{3}{8}$.

Same Numerator Strategy

- When the numerators are the same, look at the denominators to compare the size of the pieces.

Think: $\frac{1}{8}$ is smaller than $\frac{1}{4}$ because there are more pieces.

$$\frac{3}{} < \frac{3}{}$$

So, the _____ weigh less.

Share and Show

1. Compare $\frac{7}{8}$ and $\frac{5}{6}$.

   Think: What is missing from each whole?

   Write <, >, or =. $\frac{7}{8}$ ◯ $\frac{5}{6}$

Compare. Write <, >, or =. Write the strategy you used.

2. $\frac{1}{2}$ ◯ $\frac{2}{3}$

   _____

3. $\frac{3}{4}$ ◯ $\frac{2}{4}$

   _____

4. $\frac{3}{8}$ ◯ $\frac{3}{6}$

   _____

5. $\frac{3}{4}$ ◯ $\frac{7}{8}$

   _____

MATHEMATICAL PRACTICES ①

Make Sense of Problems
How do the missing pieces in Exercise 1 help you compare $\frac{7}{8}$ and $\frac{5}{6}$?

Name _____

Compare. Write <, >, or =. Write the strategy you used.

6. $\frac{1}{2}$ ◯ $\frac{2}{2}$

_____

7. $\frac{1}{3}$ ◯ $\frac{1}{4}$

_____

8. $\frac{2}{3}$ ◯ $\frac{5}{6}$

_____

9. $\frac{4}{6}$ ◯ $\frac{4}{2}$

_____

Name a fraction that is less than or greater than the given fraction. Draw to justify your answer.

10. less than $\frac{5}{6}$ _____

11. greater than $\frac{3}{8}$ _____

12. **GO DEEPER** Luke, Seth, and Anja have empty glasses. Mr. Gabel pours $\frac{3}{6}$ cup of orange juice in Seth's glass. Then he pours $\frac{1}{6}$ cup of orange juice in Luke's glass and $\frac{2}{6}$ cup of orange juice in Anja's glass. Who gets the most orange juice?

_____

13. **THINK SMARTER** **What's the Error?** Jack says that $\frac{5}{8}$ is greater than $\frac{5}{6}$ because the denominator 8 is greater than the denominator 6. Describe Jack's error. Draw a picture to explain your answer.

_____

_____

_____

Unlock the Problem

14. **MATHEMATICAL PRACTICE ①** **Analyze** Tracy is making blueberry muffins. She is using $\frac{4}{4}$ cup of honey and $\frac{4}{2}$ cups of flour. Does Tracy use more honey or more flour?

a. What do you need to know?

_____

b. What strategy will you use to compare the fractions?

_____

c. Show the steps you used to solve the problem.

d. Complete the comparison.

▭ > ▭

So, Tracy uses more _____.

---

15. **THINK SMARTER** Compare the fractions. Circle a symbol that makes the statement true.

$\frac{2}{8}$ $\begin{array}{c} > \\ < \\ = \end{array}$ $\frac{2}{4}$ $\frac{1}{4}$ $\begin{array}{c} > \\ < \\ = \end{array}$ $\frac{4}{8}$

Compare Fractions

Common Core **COMMON CORE STANDARD—3.NF.A.3d**
Develop an understanding of fractions as numbers.

Compare. Write <, >, or =. Write the strategy you used.

1. $\frac{3}{8}$ $\boxed{<}$ $\frac{3}{4}$

   Think: The numerators are the same. Compare the denominators. The greater fraction will have the lesser denominator.

   same numerator _____

2. $\frac{2}{3}$ $\bigcirc$ $\frac{7}{8}$

3. $\frac{3}{4}$ $\bigcirc$ $\frac{1}{4}$

Name a fraction that is less than or greater than the given fraction. Draw to justify your answer.

4. greater than $\frac{1}{3}$ —

5. less than $\frac{3}{4}$ —

Problem Solving

6. At the third-grade party, two groups each had their own pizza. The blue group ate $\frac{7}{8}$ pizza. The green group ate $\frac{2}{8}$ pizza. Which group ate more of their pizza?

7. Ben and Antonio both take the same bus to school. Ben's ride is $\frac{7}{8}$ mile. Antonio's ride is $\frac{3}{4}$ mile. Who has a longer bus ride?

8. **WRITE** ▸Math Explain how to use the missing pieces strategy to compare two fractions. Include a diagram with your explanation.

Lesson Check (3.NF.A.3d)

1. Compare $\frac{2}{3}$ and $\frac{7}{8}$. Write $<$, $>$, or $=$.

$\frac{2}{3} \bigcirc \frac{7}{8}$

2. What symbol makes the statement true? Write $<$, $>$, or $=$.

$\frac{2}{4} \bigcirc \frac{2}{6}$

Spiral Review (3.OA.A.4, 3.NBT.A.3, 3.NF.A.3c)

3. Cam, Stella, and Rose each picked 40 apples. They put all their apples in one crate. How many apples are in the crate?

4. Each shape is 1 whole. What fraction is represented by the shaded part of the model?

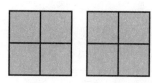

5. What related multiplication fact can you use to find $16 \div \blacksquare = 2$?

6. What is the unknown factor?

$9 \times \blacksquare = 36$

FOR MORE PRACTICE
GO TO THE
Personal Math Trainer

Name _____

 **Mid-Chapter Checkpoint**

Concepts and Skills

1. When two fractions refer to the same whole, explain why the fraction with a lesser denominator has larger pieces than the fraction with a greater denominator. (3.NF.A.3d)

_____

_____

_____

2. When two fractions refer to the same whole and have the same denominators, explain why you can compare only the numerators. (3.NF.A.3d)

_____

_____

_____

Compare. Write <, >, or =. (3.NF.A.3d)

3. $\frac{1}{6}$ ◯ $\frac{1}{4}$ 4. $\frac{1}{8}$ ◯ $\frac{1}{8}$ 5. $\frac{2}{8}$ ◯ $\frac{2}{3}$

6. $\frac{4}{2}$ ◯ $\frac{1}{2}$ 7. $\frac{7}{8}$ ◯ $\frac{3}{8}$ 8. $\frac{5}{6}$ ◯ $\frac{2}{3}$

9. $\frac{2}{4}$ ◯ $\frac{3}{4}$ 10. $\frac{6}{6}$ ◯ $\frac{6}{8}$ 11. $\frac{3}{4}$ ◯ $\frac{7}{8}$

Name a fraction that is less than or greater than the given fraction. Draw to justify your answer. (3.NF.A.3d)

12. greater than $\frac{2}{6}$ _____   13. less than $\frac{2}{3}$ _____

14. Two walls in Tiffany's room are the same size. Tiffany paints $\frac{1}{4}$ of one wall. Roberto paints $\frac{1}{8}$ of the other wall. Who painted a greater amount in Tiffany's room? (3.NF.A.3d)

_____

15. Matthew ran $\frac{5}{8}$ mile during track practice. Pablo ran $\frac{5}{6}$ mile. Who ran farther? (3.NF.A.3d)

_____

16. Mallory bought 6 roses for her mother. Two-sixths of the roses are red and $\frac{4}{6}$ are yellow. Did Mallory buy fewer red roses or yellow roses? (3.NF.A.3d)

_____

17. **GO DEEPER** Lani used $\frac{2}{3}$ cup of raisins, $\frac{3}{8}$ cup of cranberries, and $\frac{3}{4}$ cup of oatmeal to bake cookies. Which ingredient did Lani use the least amount of? (3.NF.A.3d)

_____

Name _____

Compare and Order Fractions

Essential Question How can you compare and order fractions?

 Common Core Number and Operations—
Fractions—3.NF.A.3d *Also 3.NF.A.1*
MATHEMATICAL PRACTICES
MP4, MP5, MP6, MP8

Unlock the Problem

Sierra, Tad, and Dale ride their bikes to school. Sierra rides $\frac{3}{4}$ mile, Tad rides $\frac{3}{8}$ mile, and Dale rides $\frac{3}{6}$ mile. Compare and order the distances from least to greatest.

- Circle the fractions you need to use.
- Underline the sentence that tells you what you need to do.

Activity 1 Order fractions with the same numerator.

Materials ■ color pencil

You can order fractions by reasoning about the size of unit fractions.

1			
$\frac{1}{4}$	$\frac{1}{4}$	$\frac{1}{4}$	$\frac{1}{4}$

$\frac{1}{8}$	$\frac{1}{8}$	$\frac{1}{8}$	$\frac{1}{8}$	$\frac{1}{8}$	$\frac{1}{8}$	$\frac{1}{8}$	$\frac{1}{8}$

$\frac{1}{6}$	$\frac{1}{6}$	$\frac{1}{6}$	$\frac{1}{6}$	$\frac{1}{6}$	$\frac{1}{6}$

Remember
- The *more* pieces a whole is divided into, the smaller the pieces are.
- The *fewer* pieces a whole is divided into, the larger the pieces are.

STEP 1 Shade one unit fraction for each fraction strip.

_____ is the longest unit fraction.

_____ is the shortest unit fraction.

STEP 2 Shade one more unit fraction for each fraction strip.

Are the shaded fourths still the longest? _____

Are the shaded eighths still the shortest? _____

STEP 3 Continue shading the fraction strips so that three unit fractions are shaded for each strip.

Are the shaded fourths still the longest? _____

Are the shaded eighths still the shortest? _____

$\frac{3}{4}$ mile is the _____ distance. $\frac{3}{8}$ mile is the _____ distance. $\frac{3}{6}$ mile is *between* the other two distances.

So, the distances in order from least to greatest are

_____ mile, _____ mile, _____ mile.

Try This! Order $\frac{2}{6}$, $\frac{2}{3}$, and $\frac{2}{4}$ from greatest to least.

Order the fractions $\frac{2}{6}$, $\frac{2}{3}$, and $\frac{2}{4}$ by thinking about the length of the unit fraction strip. Then label the fractions *shortest*, *between*, or *longest*.

Fraction	Unit Fraction	Length
$\frac{2}{6}$		
$\frac{2}{3}$		
$\frac{2}{4}$		

Math Talk

Generalize When ordering three fractions, what do you know about the third fraction when you know which fraction is the shortest and which fraction is the longest? Explain your answer.

- When the numerators are the same, think about the

  _____ of the pieces to compare and order fractions.

So, the order from greatest to least is _____ , _____ , _____ .

Hands On

🔑 Activity 2 Order fractions with the same denominator.

Materials ■ color pencil

Shade fraction strips to order $\frac{5}{8}$, $\frac{8}{8}$, and $\frac{3}{8}$ from least to greatest.

1

| $\frac{1}{8}$ | $\frac{1}{8}$ | $\frac{1}{8}$ | $\frac{1}{8}$ | $\frac{1}{8}$ | $\frac{1}{8}$ | $\frac{1}{8}$ | $\frac{1}{8}$ | Shade $\frac{5}{8}$. |

| $\frac{1}{8}$ | $\frac{1}{8}$ | $\frac{1}{8}$ | $\frac{1}{8}$ | $\frac{1}{8}$ | $\frac{1}{8}$ | $\frac{1}{8}$ | $\frac{1}{8}$ | Shade $\frac{8}{8}$. |

| $\frac{1}{8}$ | $\frac{1}{8}$ | $\frac{1}{8}$ | $\frac{1}{8}$ | $\frac{1}{8}$ | $\frac{1}{8}$ | $\frac{1}{8}$ | $\frac{1}{8}$ | Shade $\frac{3}{8}$. |

- When the denominators are the same, the size of the pieces is the _____ .

  So, think about the _____ of pieces to compare and order fractions.

  _____ is the shortest. _____ is the longest.

  _____ is between the other two fractions.

So, the order from least to greatest is _____ , _____ , _____ .

Name _____

1. Shade the fraction strips to order $\frac{4}{6}$, $\frac{4}{4}$, and $\frac{4}{8}$ from least to greatest.

Math Talk

MATHEMATICAL PRACTICES ⑤

Use a Concrete Model Why does using fraction strips help you order fractions with unlike denominators?

1					
$\frac{1}{6}$	$\frac{1}{6}$	$\frac{1}{6}$	$\frac{1}{6}$	$\frac{1}{6}$	$\frac{1}{6}$

$\frac{1}{4}$	$\frac{1}{4}$	$\frac{1}{4}$	$\frac{1}{4}$

$\frac{1}{8}$	$\frac{1}{8}$	$\frac{1}{8}$	$\frac{1}{8}$	$\frac{1}{8}$	$\frac{1}{8}$	$\frac{1}{8}$	$\frac{1}{8}$

_____ is the shortest. _____ is the longest.

_____ is between the other two lengths. _____ , _____ , _____

Write the fractions in order from least to greatest.

✓ 2. $\frac{1}{2}$, $\frac{0}{2}$, $\frac{2}{2}$ _____ , _____ , _____

✓ 3. $\frac{1}{6}$, $\frac{1}{2}$, $\frac{1}{3}$ _____ , _____ , _____

On Your Own

Write the fractions in order from greatest to least.

4. $\frac{6}{6}$, $\frac{2}{6}$, $\frac{5}{6}$ _____ , _____ , _____

5. $\frac{1}{8}$, $\frac{1}{4}$, $\frac{1}{2}$ _____ , _____ , _____

Write the fractions in order from least to greatest.

6. THINK SMARTER
   $\frac{6}{3}$, $\frac{6}{2}$, $\frac{6}{8}$ _____ , _____ , _____

7. THINK SMARTER
   $\frac{4}{2}$, $\frac{2}{2}$, $\frac{8}{2}$ _____ , _____ , _____

8. MATHEMATICAL PRACTICE ⑥ **Compare** Pam is making biscuits.
   She needs $\frac{2}{6}$ cup of oil, $\frac{2}{3}$ cup of water, and $\frac{2}{4}$ cup of milk.
   Write the ingredients from greatest to least amount.

   _____ , _____ , _____

Problem Solving • Applications (Real World)

9. In fifteen minutes, Greg's sailboat went $\frac{3}{6}$ mile, Gina's sailboat went $\frac{6}{6}$ mile, and Stuart's sailboat went $\frac{4}{6}$ mile. Whose sailboat went the longest distance in fifteen minutes?

_____

Whose sailboat went the shortest distance?

_____

10. **GO DEEPER**   Look back at Problem 9. Write a similar problem by changing the fraction of a mile each sailboat traveled, so the answers are different from Problem 9. Then solve the problem.

WRITE ▸ *Math* • **Show Your Work**

_____

_____

_____

_____

_____

11. **THINK SMARTER**   Tom has three pieces of wood. The length of the longest piece is $\frac{3}{4}$ foot. The length of the shortest piece is $\frac{3}{8}$ foot. What might be the length of the third piece of wood?

_____

12. **THINK SMARTER**   Jesse ran $\frac{2}{4}$ mile on Monday, $\frac{2}{3}$ mile on Tuesday, and $\frac{2}{8}$ mile on Wednesday. Order the fractions from least to greatest.

$\frac{2}{4}$, $\frac{2}{3}$ and $\frac{2}{8}$ ☐ ☐ ☐

Compare and Order Fractions

Common Core   **COMMON CORE STANDARD—3.NF.A.3d**
Develop understanding of fractions as numbers.

Write the fractions in order from greatest to least.

1. $\frac{4}{4}, \frac{1}{4}, \frac{3}{4}$ $\frac{4}{4}$, $\frac{3}{4}$, $\frac{1}{4}$

Think: The denominators are the same, so compare the numerators: $4 > 3 > 1$.

2. $\frac{2}{8}, \frac{5}{8}, \frac{1}{8}$ _____, _____, _____

3. $\frac{1}{3}, \frac{1}{6}, \frac{1}{2}$ _____, _____, _____

4. $\frac{2}{3}, \frac{2}{6}, \frac{2}{8}$ _____, _____, _____

Write the fractions in order from least to greatest.

5. $\frac{2}{4}, \frac{4}{4}, \frac{3}{4}$ _____, _____, _____

6. $\frac{4}{6}, \frac{5}{6}, \frac{2}{6}$ _____, _____, _____

Problem Solving

7. Mr. Jackson ran $\frac{7}{8}$ mile on Monday. He ran $\frac{3}{8}$ mile on Wednesday and $\frac{5}{8}$ mile on Friday. On which day did Mr. Jackson run the shortest distance?

8. Delia has three pieces of ribbon. Her red ribbon is $\frac{2}{4}$ foot long. Her green ribbon is $\frac{2}{3}$ foot long. Her yellow ribbon is $\frac{2}{6}$ foot long. She wants to use the longest piece for a project. Which color ribbon should Delia use?

_____

_____

9. **WRITE** *Math* Describe how fraction strips can help you order fractions.

_____

_____

_____

© Houghton Mifflin Harcourt Publishing Company

Lesson Check (3.NF.A.3d)

1. Write the fractions in order from least to greatest.

$$\frac{1}{8}, \frac{1}{3}, \frac{1}{6}$$

2. Write the fractions in order from greatest to least.

$$\frac{3}{6}, \frac{3}{4}, \frac{3}{8}$$

Spiral Review (3.OA.B.5, 3.NF.A.1, 3.MD.B.3)

3. What fraction of the group of cars is shaded?

4. Wendy has 6 pieces of fruit. Of these, 2 pieces are bananas. What fraction of Wendy's fruit is bananas?

5. Toby collects data and makes a bar graph about his classmates' pets. He finds that 9 classmates have dogs, 2 classmates have fish, 6 classmates have cats, and 3 classmates have gerbils. What pet will have the longest bar on the bar graph?

6. The number sentence is an example of which multiplication property?

$$6 \times 7 = (6 \times 5) + (6 \times 2)$$

© Houghton Mifflin Harcourt Publishing Company

FOR MORE PRACTICE
GO TO THE
Personal Math Trainer

Name _____

Model Equivalent Fractions

Essential Question How can you use models to find equivalent fractions?

Common Core Number and Operations—Fractions—
3.NF.A.3a *Also 3.NF.A.1, 3.NF.A.2a, 3.NF.A.2b, 3.NF.A.3, 3.NF.A.3b, 3.NF.A.3c, 3.G.A.2*
MATHEMATICAL PRACTICES
MP2, MP6, MP7

Investigate

Hands On

Materials ■ sheet of paper ■ crayon or color pencil

Two or more fractions that name the same amount are called **equivalent fractions**. You can use a sheet of paper to model fractions equivalent to $\frac{1}{2}$.

A. First, fold a sheet of paper into two equal parts. Open the paper and count the parts.

There are _____ equal parts. Each part is _____ of the paper.

Shade one of the halves. Write $\frac{1}{2}$ on each of the halves.

B. Next, fold the paper in half two times. Open the paper.

Now there are _____ equal parts. Each part is

_____ of the paper.

Write $\frac{1}{4}$ on each of the fourths.

Look at the shaded parts. $\frac{1}{2} = \frac{}{4}$

C. Last, fold the paper in half three times.

Now there are _____ equal parts. Each part is

_____ of the paper.

Write $\frac{1}{8}$ on each of the eighths.

Find the fractions equivalent to $\frac{1}{2}$ on your paper.

So, $\frac{1}{2}$, ——, and —— are equivalent.

Draw Conclusions

1. Explain how many $\frac{1}{8}$ parts are equivalent to one $\frac{1}{4}$ part on your paper.

_____

_____

2. **THINK SMARTER** What do you notice about how the numerators changed for the shaded part as you folded the paper? _____

What does this tell you about the change in the number of parts? _____

How did the denominators change for the shaded part as you folded? _____

What does this tell you about the change in the size of the parts? _____

Make Connections

You can use a number line to find equivalent fractions.

Find a fraction equivalent to $\frac{2}{3}$.

Materials ■ fraction strips

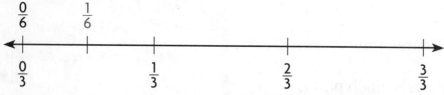

STEP 1 Draw a point on the number line to represent the distance $\frac{2}{3}$.

_____

STEP 2 Use fraction strips to divide the number line into sixths. At the end of each strip, draw a mark on the number line and label the marks to show sixths.

_____

STEP 3 Identify the fraction that names the same point as $\frac{2}{3}$. _____

So, $\frac{2}{3} = \dfrac{}{6}$.

540

Name _____

Shade the model. Then divide the pieces to find the equivalent fraction.

1.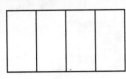

$$\frac{1}{4} = \frac{}{8}$$

2.

$$\frac{2}{3} = \frac{}{6}$$

Use the number line to find the equivalent fraction.

3.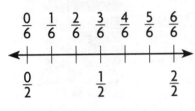

$$\frac{1}{2} = \frac{}{6}$$

4.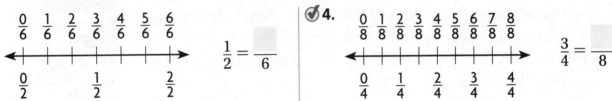

$$\frac{3}{4} = \frac{}{8}$$

Problem Solving • Applications Real World

5. MATHEMATICAL PRACTICE 6 **Explain** why $\frac{2}{2} = 1$.
Write another fraction that is equal to 1. Draw to justify your answer.

_____

_____

_____

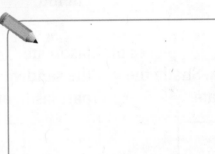

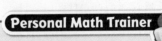
Personal Math Trainer

6. THINK SMARTER + For numbers 6a–6d, select True or False to tell whether the fractions are equivalent.

6a. $\frac{6}{6}$ and $\frac{3}{3}$ ○ True   ○ False

6b. $\frac{4}{6}$ and $\frac{1}{3}$ ○ True   ○ False

6c. $\frac{2}{3}$ and $\frac{3}{6}$ ○ True   ○ False

6d. $\frac{1}{3}$ and $\frac{2}{6}$ ○ True   ○ False

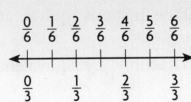

Connect to Reading

Summarize

You can *summarize* the information in a problem by underlining it or writing the information needed to answer a question.

Read the problem. Underline the important information.

7. **THINK SMARTER** Mrs. Akers bought three sandwiches that were the same size. She cut the first one into thirds. She cut the second one into fourths and the third one into sixths. Marian ate 2 pieces of the first sandwich. Jason ate 2 pieces of the second sandwich. Marcos ate 3 pieces of the third sandwich. Which children ate the same amount of a sandwich? Explain.

The first sandwich was cut into _____.	The second sandwich was cut into _____.	The third sandwich was cut into _____.
Marian ate _____ pieces of the sandwich. Shade the part Marian ate.	Jason ate _____ pieces of the sandwich. Shade the part Jason ate.	Marcos ate _____ pieces of the sandwich. Shade the part Marcos ate.
Marian ate ── of the first sandwich.	Jason ate ── of the second sandwich.	Marcos ate ── of the third sandwich.

Are all the fractions equivalent? _____

Which fractions are equivalent? ── = ──

So, _____ and _____ ate the same amount of a sandwich.

Model Equivalent Fractions

Common Core COMMON CORE STANDARD—3.NF.A.3a
Develop understanding of fractions as numbers.

Shade the model. Then divide the pieces to find the equivalent fraction.

1.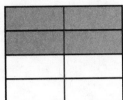

$$\frac{2}{4} = \frac{4}{8}$$

Use the number line to find the equivalent fraction.

2. $\frac{0}{8}$ $\frac{1}{8}$ $\frac{2}{8}$ $\frac{3}{8}$ $\frac{4}{8}$ $\frac{5}{8}$ $\frac{6}{8}$ $\frac{7}{8}$ $\frac{8}{8}$

$\frac{0}{4}$ $\frac{1}{4}$ $\frac{2}{4}$ $\frac{3}{4}$ $\frac{4}{4}$

$$\frac{3}{4} = \frac{\square}{8}$$

Problem Solving *Real World*

3. Mike says that $\frac{3}{3}$ of his fraction model is shaded blue. Ryan says that $\frac{6}{6}$ of the same model is shaded blue. Are the two fractions equivalent? If so, what is another equivalent fraction?

_____

4. Brett shaded $\frac{4}{8}$ of a sheet of notebook paper. Aisha says he shaded $\frac{1}{2}$ of the paper. Are the two fractions equivalent? If so, what is another equivalent fraction?

_____

5. **WRITE** ▸*Math* Draw a number line that shows two equivalent fractions. Label your number line and explain how you know the fractions are equivalent.

_____

Lesson Check (3.NF.A.3b)

1. Name a fraction equivalent to $\frac{2}{3}$.

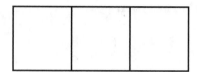

_____

2. Find the fraction equivalent to $\frac{1}{4}$.

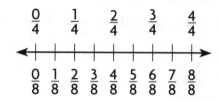

_____

Spiral Review (3.OA.A.3, 3.OA.C.7, 3.NF.A.1)

3. Eric practiced piano and guitar for a total of 8 hours this week. He practiced the piano for $\frac{1}{4}$ of that time. How many hours did Eric practice the piano this week?

_____

4. Kylee bought a pack of 12 cookies. One-third of the cookies are peanut butter. How many of the cookies in the pack are peanut butter?

_____

5. There are 56 students going to the game. The coach puts 7 students in each van. How many vans are needed to take the students to the game?

_____

6. Write a division equation for the picture.

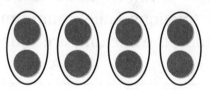

_____

_____

FOR MORE PRACTICE
GO TO THE
Personal Math Trainer

Equivalent Fractions

Essential Question How can you use models to name equivalent fractions?

Common Core
Number and Operations—Fractions—3.NF.A.3b Also 3.NF.A.1, 3.NF.A.3, 3.NF.A.3a, 3.G.A.2

MATHEMATICAL PRACTICES
MP2, MP3, MP4

Unlock the Problem

Cole brought a submarine sandwich to the picnic. He shared the sandwich equally with 3 friends. The sandwich was cut into eighths. What are two ways to describe the part of the sandwich each friend ate?

• How many people shared the sandwich?

_____

Cole grouped the smaller pieces into twos. Draw circles to show equal groups of two pieces to show what each friend ate.

There are 4 equal groups. Each group is $\frac{1}{4}$ of the whole sandwich. So, each friend ate $\frac{1}{4}$ of the whole sandwich.

How many eighths did each friend eat? _____

$\frac{1}{4}$ and _____ are equivalent fractions since they both name

the _____ amount of the sandwich.

So, $\frac{1}{4}$ and _____ of the sandwich are two ways to describe the part of the sandwich each friend ate.

Try This! Circle equal groups. Write an equivalent fraction for the shaded part of the whole.

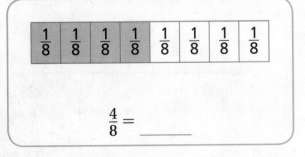

$$\frac{4}{8} = \underline{\quad\quad}$$

Math Talk

MATHEMATICAL PRACTICES ❸

Apply What is a different way you could have circled the equal groups?

🔑 Example Model the problem.

Heidi ate $\frac{3}{6}$ of her fruit bar. Molly ate $\frac{4}{8}$ of her fruit bar, which is the same size. Which girl ate more of her fruit bar?

Shade $\frac{3}{6}$ of Heidi's fruit bar and $\frac{4}{8}$ of Molly's fruit bar.

- Is $\frac{3}{6}$ greater than, less than, or equal to $\frac{4}{8}$? _____

So, both girls ate the _____ amount.

Heidi

$\frac{1}{6}$	$\frac{1}{6}$	$\frac{1}{6}$
$\frac{1}{6}$	$\frac{1}{6}$	$\frac{1}{6}$

Molly

$\frac{1}{8}$	$\frac{1}{8}$	$\frac{1}{8}$	$\frac{1}{8}$
$\frac{1}{8}$	$\frac{1}{8}$	$\frac{1}{8}$	$\frac{1}{8}$

Try This! Each shape is 1 whole. Write an equivalent fraction for the shaded part of the models.

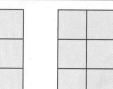

$$\frac{6}{3} = \frac{}{6}$$

Share and Show MATH BOARD

Math Talk

MATHEMATICAL PRACTICES ②

Use Reasoning Explain why equivalent fractions name the same amount.

1. Each shape is 1 whole. Use the model to find the equivalent fraction.

$$\frac{2}{4} = \frac{}{2}$$

Each shape is 1 whole. Shade the model to find the equivalent fraction.

✓ 2.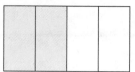

$$\frac{2}{4} = \frac{}{8}$$

3.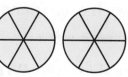

$$\frac{12}{6} = \frac{}{3}$$

4. Andy swam $\frac{8}{8}$ mile in a race. Use the number line to find a fraction that is equivalent to $\frac{8}{8}$.

$$\frac{8}{8} = \underline{}$$

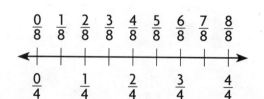

Name _____

Circle equal groups to find the equivalent fraction.

✓ **5.**

$$\frac{3}{6} = \frac{}{2}$$

6.

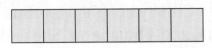

$$\frac{6}{6} = \frac{}{3}$$

On Your Own

Each shape is 1 whole. Shade the model to find the equivalent fraction.

7.

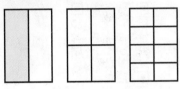

$$\frac{1}{2} = \frac{2}{} = \frac{}{8}$$

8.

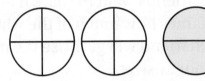

$$\frac{8}{} = \frac{4}{2}$$

Circle equal groups to find the equivalent fraction.

9.

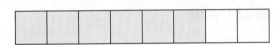

$$\frac{6}{8} = \frac{}{4}$$

10.

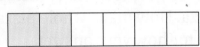

$$\frac{2}{6} = \frac{}{3}$$

11. Write the fraction that names the shaded part of each circle.

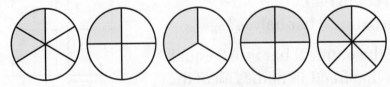

_____   _____   _____   _____   _____

Which pairs of fractions are equivalent? _____

12. **MATHEMATICAL PRACTICE ③ Apply** Matt cut his small pizza into 6 equal pieces and ate 4 of them. Josh cut his small pizza, which is the same size, into 3 equal pieces and ate 2 of them. Write fractions for the amount they each ate. Are the fractions equivalent? Draw to explain.

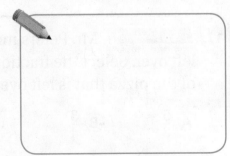

_____

Problem Solving • Applications

13. **Go DEEPER** Christy bought 8 muffins. She chose 2 apple, 2 banana, and 4 blueberry. She and her family ate the apple and banana muffins for breakfast. What fraction of the muffins did they eat? Write an equivalent fraction. Draw a picture.

_____

14. **THINK SMARTER** After dinner, $\frac{2}{3}$ of the corn bread is left. Suppose 4 friends want to share it equally. What fraction names how much of the whole pan of corn bread each friend will get? Use the model on the right. Explain your answer.

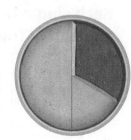

_____

_____

15. There are 16 people having lunch. Each person wants $\frac{1}{4}$ of a pizza. How many whole pizzas are needed? Draw a picture to show your answer.

_____

16. Lucy has 5 oatmeal bars, each cut in half. What fraction names all of the oatmeal bar halves? $\dfrac{}{2}$

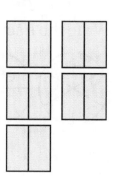

What if Lucy cuts each part of the oatmeal bar into 2 equal pieces to share with friends? What fraction names all of the oatmeal bar pieces now? $\dfrac{}{4}$

$\dfrac{}{2}$ and $\dfrac{}{4}$ are equivalent fractions.

17. **THINK SMARTER** Mr. Peters made a pizza. There is $\frac{4}{8}$ of the pizza left over. Select the fractions that are equivalent to the part of the pizza that is left over. Mark all that apply.

Ⓐ $\frac{5}{8}$ Ⓑ $\frac{3}{4}$ Ⓒ $\frac{2}{4}$ Ⓓ $\frac{1}{2}$

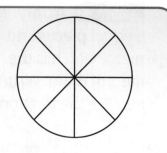

Equivalent Fractions

Common Core   COMMON CORE STANDARD—3.NF.A.3b
Develop understanding of fractions as numbers.

Each shape is 1 whole. Shade the model to find the equivalent fraction.

1.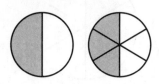

$\frac{1}{2} = \frac{3}{6}$

2.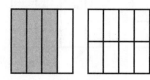

$\frac{3}{4} = \frac{6}{\boxed{}}$

Circle equal groups to find the equivalent fraction.

3.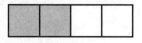

$\frac{2}{4} = \frac{\boxed{}}{2}$

4.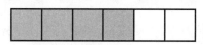

$\frac{4}{6} = \frac{\boxed{}}{3}$

Problem Solving Real World

5. May painted 4 out of 8 equal parts of a poster board blue. Jared painted 2 out of 4 equal parts of a same-size poster board red. Write fractions to show which part of the poster board each person painted.

_____

6. **WRITE** ▸ Math Explain how you can find a fraction that is equivalent to $\frac{1}{4}$.

_____

_____

Lesson Check (3.NF.A.3b)

1. What fraction is equivalent to $\frac{6}{8}$?

2. What fraction is equivalent to $\frac{1}{3}$?

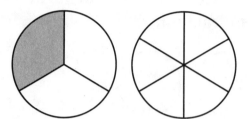

_____

_____

Spiral Review (3.OA.B.5, 3.OA.B.6, 3.OA.C.7)

3. What division number sentence is shown by the array?

_____

_____

4. Cody put 4 plates on the table. He put 1 apple on each plate. What number sentence can be used to find the total number of apples on the table?

_____

_____

5. Write a division number sentence that is a related fact to $7 \times 3 = 21$.

_____

_____

6. Find the quotient.

$$4\overline{)36}$$

_____

_____

FOR MORE PRACTICE
GO TO THE
Personal Math Trainer

✓ Chapter 9 Review/Test

1. Alexa and Rose read books that have the same number of pages. Alexa's book is divided into 8 equal chapters. Rose's book is divided into 6 equal chapters. Each girl has read 3 chapters of her book.

    Write a fraction to describe what part of the book each girl read. Then tell who read more pages. Explain.

    _____

    _____

    _____

2. David, Maria, and Simone are shading same-sized index cards for a science project. David shaded $\frac{2}{4}$ of his index card. Maria shaded $\frac{2}{8}$ of her index card and Simone shaded $\frac{2}{6}$ of her index card.

    For 2a–2d, choose Yes or No to indicate whether the comparisons are correct.

    2a. $\frac{2}{4} > \frac{2}{8}$ ○ Yes   ○ No

    2b. $\frac{2}{8} > \frac{2}{6}$ ○ Yes   ○ No

    2c. $\frac{2}{6} < \frac{2}{4}$ ○ Yes   ○ No

    2d. $\frac{2}{8} = \frac{2}{4}$ ○ Yes   ○ No

3. Dan and Miguel are working on the same homework assignment. Dan has finished $\frac{1}{4}$ of the assignment. Miguel has finished $\frac{3}{4}$ of the assignment. Which statement is correct? Mark all that apply.

    Ⓐ Miguel has completed the entire assignment.

    Ⓑ Dan has not completed the entire assignment.

    Ⓒ Miguel has finished more of the assignment than Dan.

    Ⓓ Dan and Miguel have completed equal parts of the assignment.

4. Bryan cut two peaches that were the same size for lunch. He cut one peach into fourths and the other into sixths. Bryan ate $\frac{3}{4}$ of the first peach. His brother ate $\frac{5}{6}$ of the second peach. Who ate more peach? Explain the strategy you used to solve the problem.

_____

_____

_____

5. A nature center offers 2 guided walks. The morning walk is $\frac{2}{3}$ mile. The evening walk is $\frac{3}{6}$ mile. Which walk is shorter? Explain how you can use the model to find the answer.

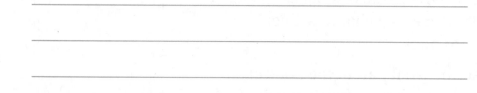

_____

_____

6. Chun lives $\frac{3}{8}$ mile from school. Gail lives $\frac{5}{8}$ mile from school.

Use the fractions and symbols to show which distance is longer.

$\frac{3}{8}$ $\frac{5}{8}$ <   >

☐ ◯ ☐

Name _____

7. **THINK SMARTER +** Mrs. Reed baked four pans of lasagna for a family party. Use the rectangles to represent the pans.

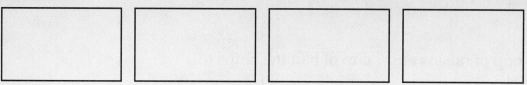

Part A

Draw lines to show how Mrs. Reed could cut one pan of lasagna into thirds, one into fourths, one into sixths, and one into eighths.

Part B

At the end of the dinner, equivalent amounts of lasagna in two pans were left. Use the models to show the lasagna that might have been left over. Write two pairs of equivalent fractions to represent the models.

---

8. Tom rode his horse for $\frac{4}{6}$ mile. Liz rode her horse for an equal distance. What is an equivalent fraction that describes how far Liz rode? Use the models to show your work.

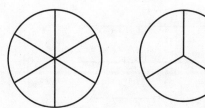

---

9. Avery prepares 2 equal-size oranges for the bats at the zoo. One dish has $\frac{3}{8}$ of an orange. Another dish has $\frac{2}{8}$ of an orange. Which dish has more orange? Show your work.

---

10. Jenna painted $\frac{1}{8}$ of one side of a fence. Mark painted $\frac{1}{6}$ of the other side of the same fence. Use >, =, or < to compare the parts that they painted.

_____

11. Bill used $\frac{1}{3}$ cup of raisins and $\frac{2}{3}$ cup of banana chips to make a snack.

For 11a–11d, select True or False for each comparison.

11a. $\frac{1}{3} > \frac{2}{3}$ ○ True   ○ False

11b. $\frac{2}{3} = \frac{1}{3}$ ○ True   ○ False

11c. $\frac{1}{3} < \frac{2}{3}$ ○ True   ○ False

11d. $\frac{2}{3} > \frac{1}{3}$ ○ True   ○ False

12. **GO DEEPER** Jorge, Lynne, and Crosby meet at the playground. Jorge lives $\frac{5}{6}$ mile from the playground. Lynne lives $\frac{4}{6}$ mile from the playground. Crosby lives $\frac{7}{8}$ mile from the playground.

Part A

Who lives closer to the playground, Jorge or Lynne? Explain how you know.

_____

_____

_____

Part B

Who lives closer to the playground, Jorge or Crosby? Explain how you know.

_____

_____

_____

554

Name _____

13. Ming needs $\frac{1}{2}$ pint of red paint for an art project. He has 6 jars that have the following amounts of red paint in them. He wants to use only 1 jar of paint. Mark all of the jars of paints that Ming could use.

    (A) $\frac{2}{3}$ pint   (D) $\frac{3}{4}$ pint

    (B) $\frac{1}{4}$ pint   (E) $\frac{3}{8}$ pint

    (C) $\frac{4}{6}$ pint   (F) $\frac{2}{6}$ pint

14. There are 12 people having lunch. Each person wants $\frac{1}{3}$ of a sub sandwich. How many whole sub sandwiches are needed? Use the models to show your answer.

_____ sub sandwiches

15. Mavis mixed $\frac{2}{4}$ quart of apple juice with $\frac{1}{2}$ quart of cranberry juice. Compare the fractions. Choose the symbol that makes the statement true.

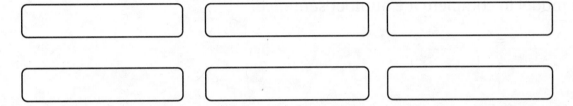

16. Pat has three pieces of fabric that measure $\frac{3}{6}$, $\frac{5}{6}$, and $\frac{2}{6}$ yards long. Write the lengths in order from least to greatest.

_____

17. Cora measures the heights of three plants. Draw a line to match each height on the left to the word on the right that describes its place in the order of heights.

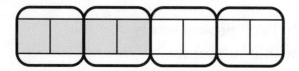

$\frac{4}{6}$ foot •   • least

$\frac{4}{4}$ foot •   • between

$\frac{4}{8}$ foot •   • greatest

18. Danielle drew a model to show equivalent fractions.

Use the model to complete the number sentence.

$\frac{1}{2} =$ _____ $=$ _____

19. Floyd caught a fish that weighed $\frac{2}{3}$ pound. Kira caught a fish that weighed $\frac{7}{8}$ pound. Whose fish weighed more? Explain the strategy you used to solve the problem.

_____

_____

_____

_____

20. Sam went for a ride on a sailboat. The ride lasted $\frac{3}{4}$ hour.

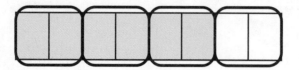

What fraction is equivalent to $\frac{3}{4}$?

_____